MONEY MATTERS 3

Written by Chris Salerno
Published by World Teachers Press®

Published with the permission of R.I.C. Publications Pty. Ltd.

Copyright © 2000 by Didax, Inc., Rowley, MA 01969. All rights reserved.

First published by R.I.C. Publications Pty. Ltd., Perth, Western Australia.

Limited reproduction permission: The publisher grants permission to individual teachers who have purchased this book to reproduce the blackline masters as needed for use with their own students. Reproduction for an entire school or school district or for commercial use is prohibited.

Printed in the United States of America.

Order Number 2-5116
ISBN 978-1-58324-038-0

G H I J K 11 10 09 08 07

395 Main Street
Rowley, MA 01969
www.didax.com

Foreword

This series of activities is aimed at reinforcing classroom and community experiences involving money. The activities are intended as a support for strategies involving play money, other teaching aids and real experiences. The blackline masters are designed to accommodate the average student and are flexible enough to meet the need of students with varying ability levels.

Challenge activities and "Did you know…?" facts have been incorporated to provide additional stimulus for the student. Aims, assessment records and answers have also been provided for ease of programming and recording. Money games and clip art have been included to make preparation easier for you.

Titles in this series:
Money Matters – Grade/Book 1
Money Matters – Grade/Book 2
Money Matters – Grade/Book 3

Contents

Objectives/NCTM Standards 4	What's Left? 27
Did You Know? 5	Multiplication 28
Assessment Records 6	Dividing Money 29
Money Games 7	Shopping for Presents—1 30
Money Clip Art 8-10	Shopping for Presents—2 31
Let's Add Up!—1 11	Change Your Coins and Bills 32
Let's Add Up!—2 12	Sharing 33
Fresh Fruits and Vegetables 13	Menu 34
What Price? 14	The $1.00 Bill 35
Shopping Time 15	The $5.00 Bill 36
Going Shopping 16	The $10.00 Bill 37
Shopping Lists 17	The $20.00 Bill 38
Let's Count—1 18	The $50.00 Bill 39
Let's Count—2 19	The $100.00 Bill 40
Dividing Money 20	Sharing Money 41
Ballooning! 21	Exchanging 42
Treasure! 22	Multiply and Divide—1 43
All Cents! 23	Multiply and Divide—2 44
Let's Add! 24	Multiply and Divide—3 45
Counting Your Money 25	Answers 46-47
Addition 26	*Math through Language, Grades 3-4* Sample Page 48

Objectives and NCTM Standards

Page	Title	Objective	NCTM Standard/Grades K-4
11	Let's Add Up!—1	Adding mixed sets of coins and writing the total.	Standards 4, 7
12	Let's Add Up!—2	Adding mixed sets of coins and bills and writing the total.	Standards 4, 7
13	Fresh Fruits and Vegetables	Identifying and writing the prices for given articles.	Standards 2, 7
14	What Price?	Adding coins to make a total.	Standard 2
15	Shopping Time	Identifying and writing the prices for given articles.	Standards 7, 8
16	Going Shopping	Counting money to equal a given price.	Standards 7, 8
17	Shopping Lists	Adding the prices of articles to make up a total.	Standards 7, 8
18	Let's Count—1	Counting up to $2.35 using coins and dollar bills.	Standards 6, 7, 8
19	Let's Count—2	Counting up to $5.45 using coins and dollar bills.	Standards 6, 7, 8
20	Dividing Money	Dividing small amounts. Maximum $2.50.	Standards 7, 8
21	Ballooning!	Sharing money among people. Maximum $2.00 among four.	Standards 7, 8
22	Treasure!	Adding mixed sets of coins and bills and writing the totals.	Standards 2, 7, 8
23	All Cents!	Adding amounts of money. Two addends, two digits. Maximum $1.00.	Standards 7, 8
24	Let's Add!	Adding amounts of money. Three addends. Mixed. Maximum $56.00.	Standards 7, 8
25	Counting Your Money	Adding amounts of money. Three addends. One digit. Maximum $18.00.	Standards 7, 8
26	Addition	Adding amounts of money. Whole dollars. Mixed.	Standards 7, 8
27	What's Left?	Subtracting amounts of money. Mixed.	Standards 7, 8
28	Multiplication	Multiplying amounts of money. Whole dollars. Mixed.	Standards 7, 8
29	Dividing Money	Dividing amounts of money. Whole dollars. Mixed.	Standards 7, 8
30	Shopping For Presents—1	Adding amounts to a given total.	Standards 2, 7, 8
31	Shopping For Presents—2	Adding amounts to a given total.	Standards 2, 7, 8
32	Change Your Coins & Bills	Making a set amount with a set number of coins and bills.	Standards 6, 7
33	Sharing	Sharing a set amount of money.	Standards 7, 8
34	Menu	Adding prices on a menu.	Standards 7, 8
35	The $1.00 Bill	Recognizing and describing the $1.00 bill.	Standards 2, 7
36	The $5.00 Bill	Recognizing and describing the $5.00 bill.	Standards 2, 7
37	The $10.00 Bill	Recognizing and describing the $10.00 bill.	Standard 2
38	The $20.00 Bill	Recognizing and describing the $20.00 bill.	Standard 2
39	The $50.00 Bill	Recognizing and describing the $50.00 bill.	Standards 2, 7
40	The $100.00 Bill	Recognizing and describing the $100.00 bill.	Standards 2, 7
41	Sharing Money	Sharing monetary amounts equally.	Standards 2, 7
42	Exchanging	Exchanging coins and bills into their simplest denomination.	Standards 2, 7
43	Multiply and Divide—1	Multiplication and division of algorithms. Mixed.	Standards 7, 8
44	Multiply and Divide—2	Multiplication and division of algorithms. Mixed.	Standards 7, 8
45	Multiply and Divide—3	Multiplication and division of algorithms. Mixed.	Standards 7, 8

Did You Know?

Did you know the Spanish dollar was called a "piece of eight" because it could be chopped into eight pieces?

Did you know that money is anything that is accepted by people in exchange for the things they sell or the work they do?

Did you know that today's money is mainly bills, coins made of different metals and checking accounts?

Did you know the dollar in Poland is called a "zloty"?

Did you know early people had no common system of money?

Did you know money developed because people realized that certain goods were always accepted in exchange for goods and services?

Did you know in Thailand the dollar is called a "baht"?

Did you know early coins were often irregular shapes?

Did you know that playing cards were used for money in Canada during the 1600s and 1700s?

Did you know feathers, leather, yarn and vodka were once used as forms of money?

Did you know a "miser" is someone who lives meagerly in order to save and store money?

Did you know the word "pay" comes from the Latin "pacare" meaning to make peace?

Did you know gold and silver have always been accepted as a form of money, because people realize their value?

Did you know the dollar in Sri Lanka is called a "rupee"?

Did you know a "mint" is a place where money is made by the government?

Did you know the Chinese made miniature versions of tools in metal as their first coins?

Did you know Marco Polo discovered that the Chinese were using paper money when he traveled to China in the 1200s?

Did you know the word "purse" comes from the Latin term "bursa"—meaning bag?

Did you know the dollar in Peru is called an "inti"?

Did you know the world's largest bank is 4.7 m long, 2.64 m tall and 6.52 m in circumference?

Did you know the Australian $10.00 note was the first polymer note ever produced? Polymer is a special kind of plastic used to try to stop people making copies of money.

Did you know amber and beads were once used as a primitive form of money to exchange for goods?

Did you know that cattle was seen as a very important article to trade before money was commonly used?

Did you know more valuable bills last longer than less valuable ones? This is because fewer people handle the more valuable bills.

Did you know the value of money is decided by people called "economists," who decide the quantity of goods and services that the money will buy?

Did you know the dollar in Zambia is called the "kwacha"?

Did you know coins used to be weighed to check their value? Designs were then stamped on coins to save people the trouble of weighing them to check their value.

Assessment Record

Name: _____ Date: _____

Objective:

Teacher Assessment | 1 | 2 | 3 | 4 | 5 |
Well developed   Needs to be developed

Activities _____
_____
_____

Teacher Comment _____
_____

Parent Signature/
Comment

Assessment Record

Name: _____ Date: _____

Objective:

Teacher Assessment | 1 | 2 | 3 | 4 | 5 |
Well developed   Needs to be developed

Activities _____
_____
_____

Teacher Comment _____
_____

Parent Signature/
Comment

Money Games

Title:	**Money Dominoes**
Aim:	To show that money can be represented in different ways.
Materials:	Dominoes made from cardboard.
Directions:	Follow ordinary domino rules. The first player places a domino and the other players take turns to build onto the domino chain wherever possible.

Title:	**Adding and Making Change**
Aim:	To practice addition of money and making change.
Materials:	Plastic coins, laminated money cards.
Directions:	Students place the correct plastic money on the appropriate coin spaces. Students add the coins and write the answer. Students remove the amounts stated on the money card. Students add the coins and write the amount. The money left on the card is the change. Students also need to write this amount.

Title:	**Classroom Store**
Aim:	To practice purchasing items and making change.
Materials:	Plastic coins and items with prices marked for purchase.
Directions:	Students take turns being the storeperson and the customer. The customer selects an item from the store and is responsible for giving the correct money to the salesperson. If change is needed, the salesperson must make the correct change.

Title:	**Money Snap**
Aim:	To become familiar with money.
Materials:	Snap cards made from cardboard.
Directions:	Follow ordinary Snap rules. As students add cards to the central pile, if two cards match, the student says "snap" and collects the cards.

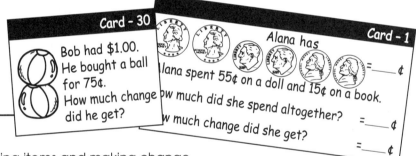

Title:	**Throwing Money**
Aim:	To practice adding monetary amounts.
Materials:	Two dice with coins on each side.
Directions:	Students throw both dice and add the totals that appear on top.
Variation:	Students could also throw both dice and subtract the totals that appear on top.

Money Clip Art

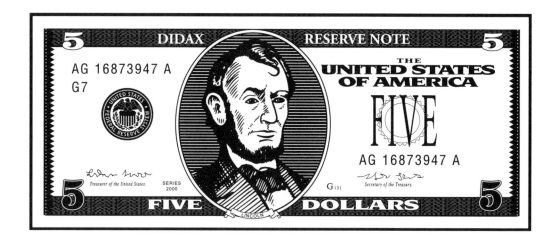

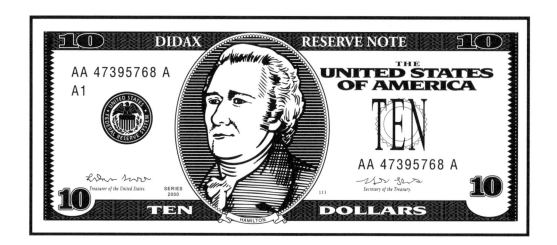

Money Clip Art

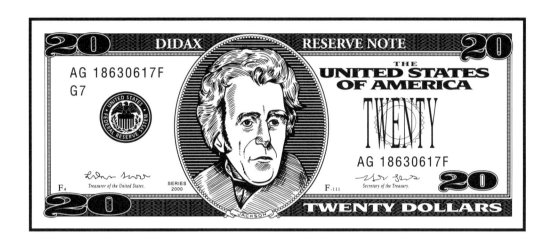

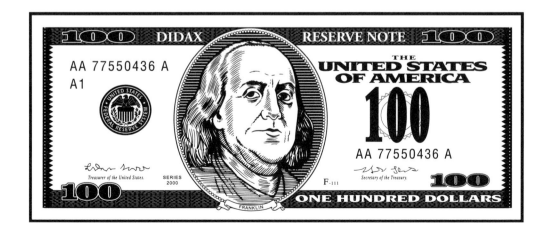

Money Clip Art

Let's Add Up!—1

Add each group of coins and write the total.

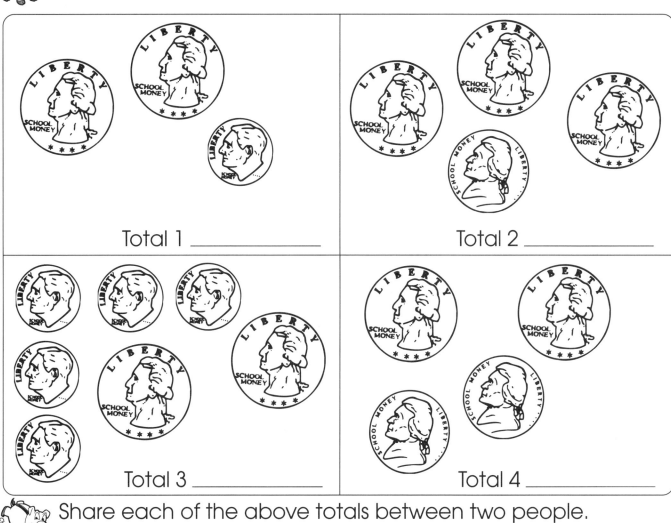

Total 1 _____

Total 2 _____

Total 3 _____

Total 4 _____

Share each of the above totals between two people. How much money would each person receive?

Total 1 _____   Total 2 _____

Total 3 _____   Total 4 _____

Explain how you solved the problems.

_____

_____

_____

_____

Let's Add Up!—2

 Add each group of coins and bills and write the total.

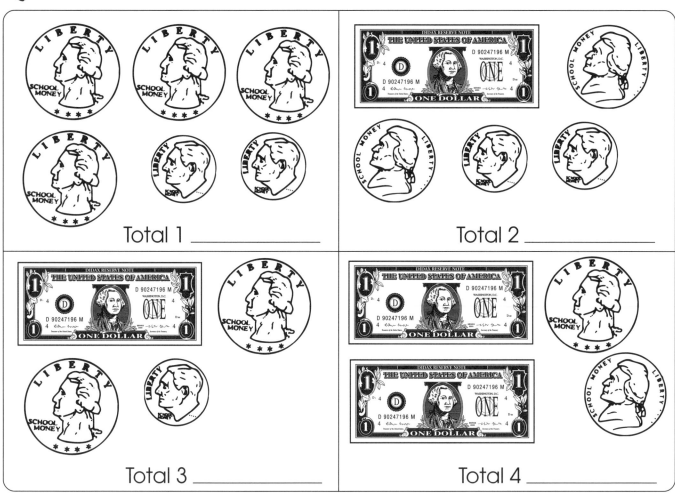

Total 1 _____ Total 2 _____

Total 3 _____ Total 4 _____

 Share each of the above totals between two people. How much money would each person receive?

Total 1 _____ Total 2 _____

Total 3 _____ Total 4 _____

Explain how you solved the problems.

_____

_____

_____

Fresh Fruits and Vegetables

In the spaces below, write the different vegetables or fruits you can see and their costs.

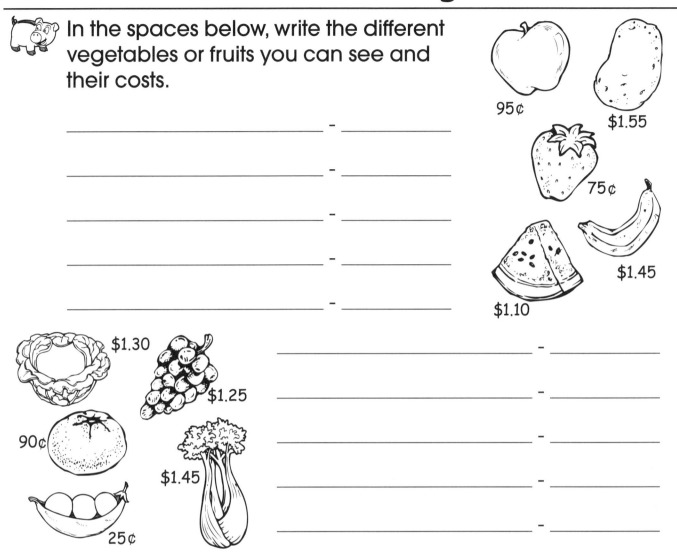

apple 95¢
potato $1.55
strawberry 75¢
banana $1.45
watermelon $1.10
cabbage $1.30
grapes $1.25
orange 90¢
celery $1.45
peas 25¢

_____ - _____

_____ - _____

_____ - _____

_____ - _____

_____ - _____

_____ - _____

_____ - _____

_____ - _____

_____ - _____

_____ - _____

Write the total cost of each of these shopping lists.

List One
2 apples _____
1 orange _____
total _____

List Two
2 strawberries _____
1 banana _____
1 apple _____
total _____

List Three
1 cabbage _____
2 oranges _____
1 potato _____
total _____

What Price?

Read the amounts on the price tags.
Make the amounts on the price tags using your money.
Make a list of the coins you use. The first one is done for you.

| $1.50 | 85¢ | 75¢ |

Coins...
25¢, 25¢
25¢, 25¢
25¢, 25¢

| $1.30 | $1.75 | $1.80 |

| $1.25 | $1.90 | $1.95 |

Using only four coins, draw the amount of $1.00.

Shopping Time

Write the prices of the items in the spaces provided.

paint set _____   bat and ball _____

marbles _____   book _____

chocolates _____   drum _____

doll _____   toy car _____

Write how much it would cost to buy:

a paint set and a bat and ball set.

_____ + _____ = _____

a doll and a book.

_____ + _____ = _____

a drum and a bag of marbles.

_____ + _____ = _____

a toy car and chocolate.

_____ + _____ = _____

Going Shopping

Count the money you would need to buy each of the items below. Write the name and cost of each item.

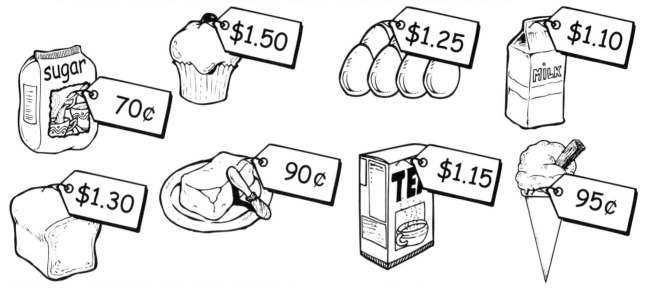

Item	Amount
Total	

Item	Amount
Total	

Solve these.

If one carton of milk costs $1.10, how much would three cartons cost? _____

If one cupcake costs $1.50, how much would three cupcakes cost? _____

If one ice-cream cone costs 95¢, how much would three ice-cream cones cost? _____

Shopping Lists

 Add up the cost of each shopping list.

pears	50¢	plums	45¢
bananas	85¢	apricots	60¢
cherries	$1.00	lemons	20¢
strawberries	65¢	oranges	25¢
peaches	$1.20	apples	$1.25

List One
bananas _____
cherries _____
total _____

List Two
apricots _____
bananas _____
total _____

List Three
apricots _____
cherries _____
total _____

List Four
apples _____
pears _____
total _____

List Five
lemons _____
bananas _____
total _____

List Six
peaches _____
strawberries _____
total _____

 If you had $2.00, how much change would you receive from each list?

List One $2.00 – _____ = _____

List Two $2.00 – _____ = _____

List Three $2.00 – _____ = _____

List Four $2.00 – _____ = _____

List Five $2.00 – _____ = _____

List Six $2.00 – _____ = _____

Let's Count—1

Use nickels, dimes, quarters and $1.00 bills to make the amounts in the boxes. Write which coins and bills you used.

35¢

For example...

3 dimes
1 nickel

95¢

75¢

$1.80

$1.95

$2.35

Write four different ways to make $2.00.

_____ = $2.00   _____ = $2.00

_____ = $2.00   _____ = $2.00

Let's Count—2

 Use nickels, dimes, quarters and $1.00 bills to make the amounts in the boxes. Write which coins and bills you used.

For example...

1 $1.00 bill
2 dimes

Add these amounts to find the totals.

50¢ + 50¢ + 20¢ = _____   20¢ + 40¢ + 5¢ = _____

$1.00 + 5¢ + 10¢ = _____   15¢ + 25¢ + 5¢ = _____

Dividing Money

 Solve these division problems.

1. 2) $1.00

2. 4) $1.00

3. 2) 50¢

4. 2) $2.00

5. 3) 90¢

6. 3) $1.20

7. 5) $2.00

8. 5) $1.50

 Solve these word problems.

Jenny and Joan found $1.00 on the way to school. How much will they each get? _____

Rick was given $1.20 to buy three apples. If there was no change, what was the cost of one apple? _____

Julia and Ian had saved $2.50 over the week. If they divide their savings equally, how much would they each have? _____

Ballooning!

Share the amounts of money shown in each balloon.

Balloon 1 — 60¢ among three people

Balloon 2 — $2.00 among four people

Balloon 3 — $1.00 among five people

Balloon 4 — $1.20 among three people

Balloon 5 — 90¢ between two people

Balloon 6 — $1.80 among three people

How much does each person get?

Balloon 1 _____   Balloon 2 _____

Balloon 3 _____   Balloon 4 _____

Balloon 5 _____   Balloon 6 _____

Which coin am I? Draw the coin in the space provided.

I am round.
I am silver.
I have an eagle on my reverse.

I am a _____.

I am round.
I am silver.
I have Monticello on my reverse.

I am a _____.

© World Teachers Press®   Money Matters, Book 3

Treasure!

Count the amount of money on each island. Write your answer in the box.

Island One

Island Two

Island Three

Island Four

Island Five

Island Six

Write the total value of each coin and bill shown on this page.

nickel _____   dime _____

quarter _____ $1.00 bill _____

All Cents!

 Solve these addition problems.

1.	10¢ + 5¢	2.	15¢ + 10¢	3.	25¢ + 15¢
4.	35¢ + 20¢	5.	45¢ + 20¢	6.	35¢ + 15¢
7.	60¢ + 30¢	8.	80¢ + 15¢	9.	75¢ + 25¢
10.	50¢ + 35¢	11.	90¢ + 10¢	12.	75¢ + 15¢

 Solve this word problem.

Jenny found 10¢ on the way to school. _____¢

Rick was given 15¢ to buy an apple at recess. _____¢

Julia had saved 25¢ over the week. +_____¢

How much money did they have altogether? _____¢

Add these amounts.

5¢ + 5¢ + 5¢ + 5¢ + 5¢ + 5¢ = _____

10¢ + 10¢ + 10¢ + 10¢ + 10¢ = _____

20¢ + 20¢ + 20¢ + 20¢ + 20¢ = _____

Let's Add!

 Solve these addition problems.

1. 5¢
    5¢
  + 5¢
  _____

2. $1.00
    $6.00
  + $6.00
  _____

3. $6.00
    $3.00
  + $2.00
  _____

4. $1.00
    $3.00
  + $4.00
  _____

5. $2.00
    $3.00
  + $2.00
  _____

6. $3.00
    $5.00
  + $6.00
  _____

7.   15¢
   + 14¢
   _____

8.   15¢
   + 10¢
   _____

9. $20.00
  +$30.00
  _____

10. $22.00
  +$34.00
  _____

Which coin am I? Draw the coin in the space provided.

I am round. I am silver.
I have a president's head on my front.
I have a torch and branches on my reverse.

I am a _____ .

I am copper colored.
I have a president's head on my front.
I have the Lincoln Memorial on my reverse.

I am a _____ .

Counting Your Money

 Solve these addition problems.

1.	$2.00 $1.00 + $3.00	2.	$3.00 $4.00 + $2.00	3.	$6.00 $4.00 + $5.00
4.	$7.00 $5.00 + $1.00	5.	$3.00 $6.00 + $4.00	6.	$9.00 $1.00 + $4.00
7.	$8.00 $2.00 + $4.00	8.	$6.00 $5.00 + $7.00	9.	$4.00 $5.00 + $2.00
10.	$8.00 $4.00 + $6.00	11.	$6.00 $8.00 + $2.00	12.	$7.00 $8.00 + $3.00

 Color the cards that total $2.00.

1.	2.	3.	4.	5.
20¢ 50¢ 50¢ 5¢ 5¢	20¢ 20¢ $1.00 50¢ 10¢	50¢ 50¢ 50¢ 50¢	50¢ 50¢ 50¢ 20¢ 20¢ 10¢	$1.00 5¢ 20¢ 20¢ 20¢

Addition

 Solve these addition problems.

1. $12.00
 + $14.00
 ─────────

2. $16.00
 + $21.00
 ─────────

3. $36.00
 + $22.00
 ─────────

4. $32.00
 + $26.00
 ─────────

5. $40.00
 + $30.00
 ─────────

6. $23.00
 + $42.00
 ─────────

7. $65.00
 + $22.00
 ─────────

8. $30.00
 + $50.00
 ─────────

9. $32.00
 + $46.00
 ─────────

10. $75.00
 + $26.00
 ─────────

11. $246.00
 +$133.00
 ─────────

12. $604.00
 +$143.00
 ─────────

 Using coins and bills, show four different ways to make $2.00.

_____ = $2.00

_____ = $2.00

_____ = $2.00

_____ = $2.00

What's Left?

 Solve these subtraction problems.

1.	$6.00 − $3.00	2.	$8.00 − $5.00	3.	$9.00 − $4.00
4.	$6.00 − $2.00	5.	$9.00 − $5.00	6.	$8.00 − $2.00
7.	55¢ − 15¢	8.	35¢ − 20¢	9.	45¢ − 15¢
10.	$16.00 −$10.00	11.	$25.00 −$13.00	12.	$35.00 −$25.00

 Solve this word problem.
Rebecca began the week with $5.00.
On Tuesday, she spent 50¢ on lunch at school.
On Thursday, she caught the bus to dancing school.
The bus cost 75¢ there and 75¢ return.
On Friday, Rebecca paid $2.00 for her school field trip.
How much money did Rebecca
have left at the end of the week? _____

What do you think Rebecca should do with the rest of the money?

_____

_____

Multiplication

Solve these multiplication problems.

1.	$1.00 × 5	2.	$1.50 × 2	3.	$2.00 × 3
4.	$1.25 × 3	5.	$1.25 × 4	6.	$2.00 × 5
7.	$2.50 × 2	8.	$2.00 × 4	9.	$1.15 × 3
10.	85¢ × 2	11.	55¢ × 3	12.	$5.00 × 2

 What coin or bill would you multiply each of these numbers by to make $5.00?

5 x _____ = $5.00

500 x _____ = $5.00

50 x _____ = $5.00

20 x _____ = $5.00

Dividing Money

Solve these division problems.

1. 3) $1.20

2. 5) $2.50

3. 2) $1.50

4. 3) $1.20

5. 4) $2.00

6. 5) $5.00

7. 2) $5.00

8. 4) $1.00

Solve these word problems.

Roy and David found $5.00 on the way to the beach. How much will they each get? _____

Sharon was given $2.40 to buy three apples. If there was no change, what was the cost of one apple? _____

Simon and Brent had saved $10.00 over the year. If they divide their savings equally, how much would they each have? _____

Shopping for Presents—1

Which two gifts together would cost:

$5.00? _____ and _____

$2.50? _____ and _____

$7.20? _____ and _____

$6.50? _____ and _____

$3.60? _____ and _____

$8.50? _____ and _____

 Jessica has $5.00 to spend. How much change would she receive if she bought…

one yo-yo? $5.00 − _____ = _____

one book? $5.00 − _____ = _____

one doll? $5.00 − _____ = _____

one toy car? $5.00 − _____ = _____

one drum? $5.00 − _____ = _____

one paint set? $5.00 − _____ = _____

Shopping for Presents—2

Which two gifts together would cost:

$7.00? _____ and _____

$9.50? _____ and _____

$3.75? _____ and _____

$5.50? _____ and _____

$9.00? _____ and _____

$4.25? _____ and _____

 Michael has $10.00 to spend. How much change would he receive if he bought…

one skateboard? $10.00 − _____ = _____

one kite? $10.00 − _____ = _____

one belt? $10.00 − _____ = _____

one baby carriage? $10.00 − _____ = _____

one helmet? $10.00 − _____ = _____

one pair of in-line skates? $10.00 − _____ = _____

Change Your Coins and Bills

Write the coins and bills you would need to:

make $2.00 with two bills. _____

make $3.00 with two bills and four coins. _____

make $1.50 with one bill and two coins. _____

make $1.50 with six coins. _____

make $1.00 with four coins. _____

How many nickels in:

30¢? _____

50¢? _____

$1.00? _____

How many quarters in:

$1.00? _____

$2.00? _____

$3.00? _____

Write three different ways you can make $1.00.

_____ = $1.00

_____ = $1.00

_____ = $1.00

How many quarters are needed to make these amounts?

$1.00 _____ $1.50 _____

$2.00 _____ $2.25 _____

$3.00 _____ $3.75 _____

$4.00 _____ $5.25 _____

Sharing

 How many people can have:

50¢ if you have $4.00 to share? _____ people

30¢ if you have $1.80 to share? _____ people

40¢ if you have $1.20 to share? _____ people

60¢ if you have $2.40 to share? _____ people

70¢ if you have $1.40 to share? _____ people

 Share this money between two people.

How much would each receive? _____

 Share this money among three people.

How much would each receive? _____

 Solve these.
Ruby has $1.00 and wishes to share the money among herself and four friends.
How much money will they each get? _____

Ken has $2.00 and wishes to share the money among himself and four friends.
How much money will they each get? _____

Menu

burger $1.50	cake $1.00
soft drink $1.10	orange .... 60¢
ice-cream cone .... 50¢	apple .... 35¢
pie $1.30	milk shake $1.45

What would it cost you to buy:

a burger and a milk shake? _____

an ice-cream cone and a piece of cake? _____

a pie and an apple? _____

an apple, a burger and a soft drink? _____

a piece of pie and a milk shake? _____

an orange and a burger? _____

How much change would you receive from $2.00, if you purchased:

a burger? _____

a milk shake? _____

an apple and an orange? _____

an ice-cream cone and a soft drink? _____

Solve these.

Caroline and her friend want to share a burger. How much money will they each need to pay? _____

Chris and his friend want to share a soft drink. How much money will they each need to pay? _____

The $1.00 Bill

Our smallest bill is the $1.00 bill.

 Write the name of the person shown on the front of this bill.

_____

 Why do you think this person was chosen for the $1.00 bill?

_____

_____

What is on the reverse side of the $1.00 bill?

_____

 How many of each of the following coins are needed to make $1.00?

nickels _____   *dimes* _____

quarters _____

The $5.00 Bill

 Write the name of the person shown on the front of this bill.

_____

Why do you think this person was chosen for the $5.00 bill?

_____
_____
_____

What is on the reverse side of the $5.00 bill?

_____
_____

How many of each of the following coins or bills are needed to make $5.00?

nickels _____   dimes _____

quarters _____ $1.00 bills _____

The $10.00 Bill

 Write the name of the person shown on the front of this bill.

_____

 Why do you think this person was chosen for the $10.00 bill?

_____

_____

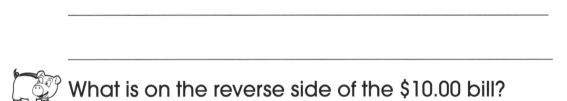 What is on the reverse side of the $10.00 bill?

_____

 Design your own $10.00 bill in the space below.

The $20.00 Bill

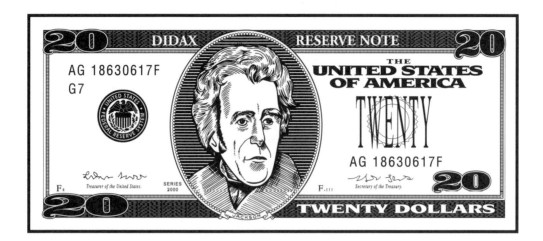

 Write the name of the person shown on the front of this bill.

_____

 Why do you think this person was chosen for the $20.00 bill?

_____

_____

_____

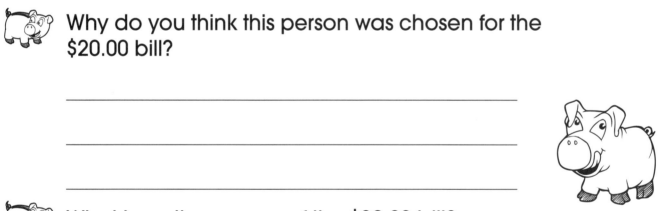

 What is on the reverse of the $20.00 bill?

_____

 How many of each of the following bills are needed to make $20.00?

$5.00 bills _____ $10.00 bills _____

Using only bills, show two different ways to make $20.00.

_____ = $20.00

_____ = $20.00

The $50.00 Bill

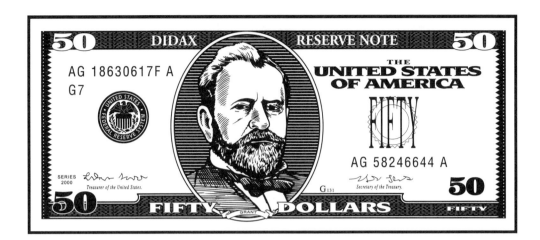

 Write the name of the person shown on the front of this bill.

_____

 Why do you think this person was chosen for the $50.00 bill?

_____

_____

_____

 What is on the reverse of the $50.00 bill?

_____

 Add these amounts. Circle the amounts that add to $50.00.

1. $35.00
   + $15.00
   _____

2. $25.00
   + $21.00
   _____

3. $30.00
   + $20.00
   _____

How can you make $50.00 using only four bills?

_____+_____+_____+_____=$50.00

The $100.00 Bill

 Write the name of the person shown on the front of this bill.
_____

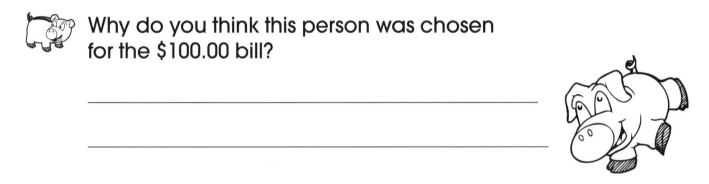

Why do you think this person was chosen for the $100.00 bill?

_____

_____

_____

 What is on the reverse of the $100.00 bill?

_____

 What do you need to add to these so they total $100.00?

1. $35.00   2. $55.00   3. $70.00
  + _____   + _____   + _____
    $100.00 $100.00 $100.00

How can you make $100.00 using only four bills?

_____+_____+_____+_____=$100.00

Sharing Money

 Share these amounts equally:

between two people.

How much money will each person get? _____

among four people.

How much money will each person get? _____

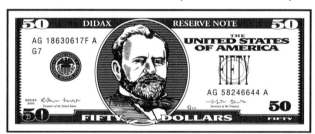

among five people.

How much money will each person get? _____

 Solve these word problems.

Chris, Brian and Les want to buy a skateboard that costs $60.00.
How much money will they each need? _____

Jenny, Simon, Brent and Anne want to buy a computer game that costs $100.00.
How much money will they each need? _____

Debra, Cathy, Tina and Ken want to buy a clubhouse that costs $280.00.
How much money will they each need? _____

Exchanging

 Exchange these bills and coins for one bill.

    =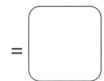

  =

 =

 Write the easiest way to make these amounts.
You need to use the least number of bills.

$25.00 = _____

$35.00 = _____

$30.00 = _____

$44.00 = _____

$56.00 = _____

$63.00 = _____

$18.00 = _____

Multiply and Divide—1

Do these problems. Your teacher will let you know if you can use coins or not.

Multiplication

1. 6 x 5¢ = _____
2. 3 x 5¢ = _____
3. 2 x 10¢ = _____
4. 5 x 5¢ = _____
5. 3 x 10¢ = _____
6. 7 x 5¢ = _____
7. 4 x 5¢ = _____
8. 6 x 10¢ = _____
9. 5 x 20¢ = _____
10. 2 x 50¢ = _____
11. 2 x $1.00 = _____
12. 3 x 50¢ = _____

Division

1. 2)20¢
2. 3)30¢
3. 4)40¢
4. 2)40¢
5. 10)60¢
6. 10)50¢
7. 10)90¢
8. 7)35¢
9. 5)45¢
10. 4)20¢
11. 5)20¢
12. 2)20¢

 Share these amounts equally.

$3.20 among four people. _____ each

$4.80 among four people. _____ each

$5.60 among four people. _____ each

Multiply and Divide—2

 Do these problems. Your teacher will let you know if you can use coins or not.

Multiplication

1. 6 x 25¢ = _____
2. 3 x 75¢ = _____
3. 2 x 50¢ = _____
4. 5 x 20¢ = _____
5. 3 x 40¢ = _____
6. 7 x 10¢ = _____
7. 4 x 50¢ = _____
8. 6 x 50¢ = _____
9. 5 x 25¢ = _____
10. 2 x $2.00 = _____
11. 2 x $1.00 = _____
12. 3 x $2.50 = _____

Division

1. 2)$1.50
2. 3)$1.80
3. 4)$1.60
4. 2)$3.00
5. 10)$5.00
6. 10)$2.50
7. 10)$3.00
8. 7)$2.10
9. 5)$10.00
10. 4)$2.00
11. 5)$3.00
12. 2)$50.00

Multiply and Divide—3

 Do these problems. Your teacher will let you know if you can use coins or not.

Multiplication

1. 6 x $1.00 = _____
2. 3 x $15.00 = _____
3. 2 x $5.00 = _____
4. 5 x $2.50 = _____
5. 3 x $2.50 = _____
6. 7 x $2.00 = _____
7. 4 x $5.00 = _____
8. 6 x $1.50 = _____
9. 5 x $10.00 = _____
10. 2 x $25.00 = _____
11. 2 x $10.00 = _____
12. 3 x $10.00 = _____

Division

1. 2)$10.00
2. 3)$18.00
3. 4)$20.00
4. 2)$50.00
5. 10)$50.00
6. 10)$100.00
7. 10)$30.00
8. 7)$21.00
9. 5)$100.00
10. 4)$24.00
11. 5)$30.00
12. 2)$30.00

Answers

11 — Let's Add Up!—1
Add each group
Total 1 – 60¢
Total 2 – 80¢
Total 3 – $1.00
Total 4 – 60¢
Share each of the above
Total 1 – 30¢
Total 2 – 40¢
Total 3 – 50¢
Total 4 – 30¢
Explain how you
Teacher check

12 — Let's Add Up!—2
Add each group
Total 1 – $1.20
Total 2 – $1.30
Total 3 – $1.60
Total 4 – $2.30
Share each of the above
Total 1 – 60¢
Total 2 – 65¢
Total 3 – 80¢
Total 4 – $1.15
Explain how you
Teacher check

13 — Fresh Fruits and Vegetables
In the space below
potato – $1.55
watermelon – $1.10
apple – 95¢
strawberry – 75¢
banana – $1.45
peas – 25¢
celery – $1.45
orange – 90¢
bunch of grapes – $1.25
cabbage – $1.30
Write the total cost
List One – $2.80
List Two – $3.90
List Three – $4.65

14 — What Price?
Read the amounts
Teacher check
Using only four
25¢, 25¢, 25¢, 25¢

15 — Shopping Time
Write the prices
paint set – $1.05
bat and ball – $1.95
marbles – $1.25
book – 90¢
chocolates – $1.45
drum – $1.50
doll – $1.65
toy car – $1.35

Write how much
$1.05 + $1.95 = $3.00
$1.65 + 90¢ = $2.55
$1.50 + $1.25 = $2.75
$1.35 + $1.45 = $2.80

16 — Going Shopping
Count the money
Teacher check
Solve these
$3.30; $4.50; $2.85

17 — Shopping Lists
Add the cost
List One – $1.85
List Two – $1.45
List Three – $1.60
List Four – $1.75
List Five – $1.05
List Six – $1.85
If I had $2.00
List One – 15¢
List Two – 55¢
List Three – 40¢
List Four – 25¢
List Five – 95¢
List Six – 15¢

18 — Let's Count—1
Teacher check

19 — Let's Count—2
Use nickels, dimes
Teacher check
Add these amounts
$1.20; 65¢; $1.15; 45¢

19 — Dividing Money
Solve these division
50¢; 25¢; 25¢; $1.00; 30¢; 40¢; 40¢; 30¢
Solve these word problems
50¢; 40¢; $1.25

21 — Ballooning!
How much does each
Balloon 1 – 20¢; Balloon 2 – 50¢;
Balloon 3 – 20¢; Balloon 4 – 40¢;
Balloon 5 – 45¢; Balloon 6 – 60¢
Which coin am I?
quarter; nickel

22 — Treasure!
Count the amounts
Island One – 85¢
Island Two – $1.35
Island Three – 50¢
Island Four – $1.50
Island Five – 95¢
Island Six – $1.60
Write the total value
nickel – 45¢; dime – 30¢;
quarter – $3.00; $1.00 bill – $3.00

23 — All Cents!
Solve these addition
1. 15¢ 2. 25¢ 3. 40¢
4. 55¢ 5. 65¢ 6. 50¢
7. 90¢ 8. 95¢ 9. $1.00
10. 85¢ 11. $1.00 12. 90¢
Solve this word
10¢ + 15¢ + 25¢ = 50¢
Add these amounts
30¢; 50¢; $1.00

24 — Let's Add!
Solve these addition
1. 15¢ 2. $13.00 3. $11.00
4. $8.00 5. $7.00 6. $14.00
7. 29¢ 8. 25¢
9. $50.00 10. $56.00
Which coin am I
dime; penny

25 — Counting Your Money
Solve these addition
1. $6.00 2. $9.00 3. $15.00
4. $13.00 5. $13.00 6. $14.00
7. $14.00 8. $18.00 9. $11.00
10. $18.00 11. $16.00 12. $18.00
Color the cards
cards 2, 3 and 4 should be colored

26 — Addition
Solve these addition
1. $26.00 2. $37.00 3. $58.00
4. $58.00 5. $70.00 6. $65.00
7. $87.00 8. $80.00 9. $78.00
10. $101.00 11. $379.00
12. $747.00
Using coins and bills
Teacher check

27 — What's Left?
Solve these subtraction
1. $3.00 2. $3.00 3. $5.00
4. $4.00 5. $4.00 6. $6.00
7. 40¢ 8. 15¢ 9. 30¢
10. $6.00 11. $12.00 12. $10.00
Solve these word
$1.00; Teacher check

28 — Multiplication
Solve these multiplication
1. $5.00 2. $3.00 3. $6.00
4. $3.75 5. $5.00 6. $10.00
7. $5.00 8. $8.00 9. $3.45
10. $1.70 11. $1.65 12. $10.00
What coin or bill
$1.00 bill; penny; dime; quarter

29 — Dividing Money
Solve these division
1. 40¢ 2. 50¢ 3. 75¢
4. 40¢ 5. 50¢ 6. $1.00
7. $2.50 8. 25¢
Solve these word
$2.50; 80¢; $5.00

Answers

30 — Shopping for Presents—1
Which two gifts
Teacher check
Jessica has $5.00
$3.60; $2.80; $2.80; $4.25; $1.50; $3.50

31 — Shopping for Presents—2
Which two gifts
Teacher check
Michael has $10.00
$2.50; $7.75; $9.00; $6.50; $8.00; $4.50

32 — Change Your Coins & Bills
Write the coins
$2.00 – two $1.00 bills
$3.00 – two $1.00 bills and four quarters
$1.50 – one $1.00 bill and two quarters
$1.50 – six quarters
$1.00 – four quarters
How many nickels
6; 10; 20
How many quarters
4; 8; 12
Write four different
Teacher check
How many quarters
4; 6;
8; 9;
12; 15;
16; 21

33 — Sharing
How many people
8; 6; 3; 4; 2
Share this money
45¢; 40¢
Solve these
20¢; 40¢

34 — Menu
What would it cost
$2.95; $1.50; $1.65; $2.95; $2.75; $2.10
How much change would
50¢; 55¢; $1.05; 40¢
Solve these
75¢; 55¢

35 — The $1.00 Bill
Write the name
George Washington
Why do you think
Teacher check
What is on the reverse
Great seal of the United States
How many of each
nickels – 20
dimes – 10
quarters – 4

36 — The $5.00 Bill
Write the name
Abraham Lincoln
Why do you think
Teacher check
What is on the reverse
The Lincoln Memorial
How many of each
nickels – 100
dimes – 50
quarters – 20
$1.00 bills – 5

37 — The $10.00 Bill
Write the name
Alexander Hamilton
Why do you think
Teacher check
What is on the reverse
US Treasury Building
Design your own
Teacher check

38 — The $20.00 Bill
Write the name
Andrew Jackson
Why do you think
Teacher check
What is on the reverse
Great seal of the United States
How many of each
$5.00 bills – 4
$10.00 bills – 2
Using only bills
Teacher check

39 — The $50.00 Bill
Write the name
Ulysses S. Grant
Why do you think
Teacher check
What is on the reverse
U.S. Capitol Building
Add these amounts
$50.00; $46.00; $50.00; 1 and 3 should be circled
How can you make
$10 bill + $10 bill + $10 bill + $20 bill = $50.00, or
$20 bill + $20 bill + $5 bill + $5 bill = $50.00

40 — The $100.00 Bill
Write the name
Ben Franklin
Why do you think
Teacher check
What is on the reverse
Independence Hall
What do you need to add
$65.00
$45.00
$30.00

How can you make
$50 bill + $20 bill + $20 bill + $10 bill = $100.00

41 — Sharing Money
Share these amounts
$5.00; $5.00; $10.00
Solve these word
$20.00; $25.00; $70.00

42 — Exchanging
Exchange these
$1.00 bill; $5.00 bill; $10.00 bill
Write the easiest way
$25.00 – $20.00, $5.00
$35.00 – $20.00, $10.00, $5.00,
$30.00 – $20.00, $10.00,
$44.00 – 2 x $20.00, 4 x $1.00
$56.00 – $50.00, $5.00, $1.00
$63.00 – $50.00, $10.00, 3 x $1.00
$18.00 – $10.00, $5.00, 3 x $1.00

43 — Multiply and Divide—1
Multiplication
1. 30¢ 2. 15¢ 3. 20¢ 4. 25¢
5. 30¢ 6. 35¢ 7. 20¢ 8. 60¢
9. $1.00 10. $1.00 11. $2.00
12. $1.50
Division
1. 10¢ 2. 10¢ 3. 10¢ 4. 20¢
5. 6¢ 6. 5¢ 7. 9¢ 8. 5¢ 9. 9¢
10. 5¢ 11. 4¢ 12. 10¢
Share these amounts…
80¢; $1.20; $1.40

44 — Multiply and Divide—2
Multiplication
1. $1.50 2. $2.25 3. $1.00 4. $1.00
5. $1.20 6. 70¢ 7. $2.00 8. $3.00
9. $1.25 10. $4.00 11. $2.00
12. $7.50
Division
1. 75¢ 2. 60¢ 3. 40¢ 4. $1.50
5. 50¢ 6. 25¢ 7. 30¢ 8. 30¢ 9. $2.00
10. 50¢ 11. 60¢ 12. $25.00

45 — Multiply and Divide—3
Multiplication
1. $6.00 2. $45.00 3. $10.00
4. $12.50 5. $7.50 6. $14.00
7. $20.00 8. $9.00 9. $50.00
10. $50.00 11. $20.00
12. $30.00
Division
1. $5.00 2. $6.00 3. $5.00 4. $25.00
5. $5.00 6. $10.00 7. $3.00 8. $3.00
9. $20.00 10. $6.00 11. $6.00
12. $15.00

A Visit to the Zoo

Sample Page from World Teachers Press' *Math through Language Grades 3-4* Book

A class of children went to the zoo. They went on May 23rd.
What day did they go on?

May

Sun	Mon	Tue	Wed	Thur	Fri	Sat
	1	2	3	4	5	6
7	8	9	10	11	12	13
14	15	16	17	18	19	20
21	22	23	24	25	26	27
28	29	30	31			

_____

The next day at school they did lots of zoo activities.
What was this day and date?

_____

This is the time they left school.

What time did they leave school? _____
It took them 30 minutes to travel to the zoo.

What time did they arrive? _____.

Show it on the clock.

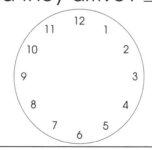

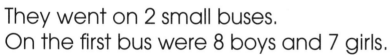

They went on 2 small buses.
On the first bus were 8 boys and 7 girls.

How many on this bus? _____
On the second bus there were 9 boys and 8 girls.

How many on this bus? _____

Which bus had more children on it? _____

How many more? _____